WOMEN SCIENTISTS

HIDDEN IN HISTORY

Cynthia O'Brien

CRABTREE
Publishing Company
www.crabtreebooks.com

Author: Cynthia O'Brien

Editorial director: Kathy Middleton

Editors: Sarah Eason, Ellen Rodger

Proofreaders: Jennifer Sanderson, Debbie Greenberg

Design: Paul Myerscough and Jessica Moon

Cover design: Emma DeBanks and Jessica Moon

Illustrations: Jessica Moon

Photo research: Rachel Blount

Production coordinator and Prepress technician: Margaret Amy Salter

Print coordinator: Katherine Berti

Written, developed, and produced by Calcium

Photo Credits:
t=Top, c=Center, b=Bottom, l= Left, r=Right

Inside: Shutterstock: Tony Baggett: p. 17; Marusya Chaika: p. 4; Everett Historical: p. 34t; Fusebulb: p. 31b; Liya Graphics: p. 18b; Marzolino: p. 11t; Muratart: p. 25; Naeblys: p. 41; Nechaevkon: p. 35; Galina Savina: p. 9; Seveniwe: p. 23; Anke van Wyk: p. 45; Wikimedia Commons: pp. 19, 36, 37, 43t; American Colony Jerusalem: p. 10; William E. Artis: p. 31t; Jessica Chapman: pp. 26l, 28; Raymond Gosling/King's College London: p. 18t; Eric Hunt: p. 33; Adam J. W. C: p. 27; JoJan: p. 21; Dorothea Lange: p. 32t; Charles Levy: p. 13t; Library of the University of Bologna: p. 6b; Los Angeles County Museum of Art: p. 6t; Maria Sibylla Merian: p. 20t; NASA/GSFC/SDO: p. 29; NSF/USAP photo by Steve Clabuesch: p. 39; Mike Peel (www.mikepeel.net). Crop by PhilipTerryGraham: p. 15; H. Raab: p. 43b; RAF-YYC from Calgary, Canada: p. 40; Stockholm00000: p. 13b; The-time-line: p. 44t; Wellcome Images: pp. 5t, 7.

Library and Archives Canada Cataloguing in Publication

Title: Women scientists : hidden in history / Cynthia O'Brien.
Names: O'Brien, Cynthia (Cynthia J.), author.
Description: Series statement: Hidden history | Includes index.
Identifiers: Canadiana (print) 20200153269 | Canadiana (ebook) 202001532677 | ISBN 9780778773009 (hardcover) | ISBN 9780778773078 (softcover) | ISBN 9781427124784 (HTML)
Subjects: LCSH: Women scientists—Biography—Juvenile literature. | LCSH: Women scientists—History—Juvenile literature.
Classification: LCC Q141 .O27 2020 | DDC j509.2/52—dc23

Library of Congress Cataloging-in-Publication Data

Names: O'Brien, Cynthia (Cynthia J.), author.
Title: Women scientists hidden in history / Cynthia O'Brien.
Description: New York, New York : Crabtree Publishing Company, [2020] | Series: Hidden history | Includes bibliographical references and index.
Identifiers: LCCN 2019054371 (print) | LCCN 2019054372 (ebook) ISBN 9780778773009 (hardcover) | ISBN 9780778773078 (paperback) | ISBN 9781427124784 (ebook)
Subjects: LCSH: Women scientists--History--Juvenile literature. | Women scientists--Biography--Juvenile literature. | Science--History--Juvenile literature.
Classification: LCC Q130 .O27 2020 (print) | LCC Q130 (ebook) | DDC 509.2/52--dc23
LC record available at https://lccn.loc.gov/2019054371
LC ebook record available at https://lccn.loc.gov/2019054372

Crabtree Publishing Company
www.crabtreebooks.com 1-800-387-7650

Printed in the U.S.A./022020/CG20200102

Published in Canada
Crabtree Publishing
616 Welland Ave.
St. Catharines, Ontario
L2M 5V6

Published in the United States
Crabtree Publishing
PMB 59051
350 Fifth Avenue, 59th Floor
New York, New York 10118

Published in the United Kingdom
Crabtree Publishing
Maritime House
Basin Road North, Hove
BN41 1WR

Published in Australia
Crabtree Publishing
Unit 3 - 5 Currumbin Court
Capalaba
QLD 4157

Contents

CHAPTER 1

HISTORY'S HIDDEN HEROINES

When you think of a scientist, do you picture a man in a white coat? If you do, you are not alone. When it comes to naming scientists, people also think of men. Albert Einstein is probably the first name to come to mind. On the other hand, many people find it difficult to name a single woman scientist! This is not because women scientists have not existed—they have been working throughout history. In fact, from ancient times to today, fearless females have made amazing discoveries in scientific fields, from astronomy **and** chemistry **to medicine and** physics.

Who Are These Women Scientists?

One of the few women scientists most people can name is Marie Curie. But often, the names of hundreds of other female scientists have been lost or forgotten. Sometimes, we may have heard their names, but we know very little about their lives. Few people know about a chemist, or a person who studies chemistry, named Maria who lived somewhere between 100 and 400 CE. She invented equipment for chemistry experiments, but that is almost all we know. Sometimes, a woman scientist's achievement is overshadowed by a man. In the 1800s, a German-British woman, Caroline Herschel, discovered eight **comets**, but still her brother, William Herschel, became more famous for his discoveries because he was a man.

Marie Curie was the first woman to win a ***Nobel Prize*** *and the first person to win it twice. Marie shared her first prize, for physics, in 1903. She won the second one on her own in 1911, for chemistry.*

Making History Now

Today, there are still many more men than women in science. Even so, women continue to be very important to science. For example, Sara Seager is a leader in the discovery of **exoplanets**. Katrin Amunts created the most detailed map of the brain ever. Two women, Jennifer Doudna and Emmanuelle Charpentier, invented groundbreaking new **genetics** technology. These women will be in tomorrow's history books, but why are history's other women scientists missing? Let's take a closer look to find out.

Caroline Herschel was the first woman to work as a paid scientist. However, she was paid as an assistant, or helper, and not considered a "real" astronomer like her brother William, pictured left.

HIDDEN HISTORY

Potions Fit for a King

Not many people know that the first recorded chemist in history was a woman. Tapputi-Belatekallim lived about 1200 BCE and was the head perfumer for the King of Babylon, an ancient city in an area that is now the Middle East. This was a big deal in ancient times, because perfume was not just for smelling good. It was used for medicine and in **religious ceremonies**. Tapputi used her chemistry skills to make the best perfumes of the time. Just one of her recipes survives today, a healing oil made for the king.

*Tapputi invented special **laboratory** equipment to make her perfumes. She used a piece of equipment called a still to boil and separate liquids into different parts.*

So, Why Can't We Read All About It?

In many **cultures** throughout history, women were leaders, healers, astronomers, and more. Many of these women were well known and respected in their own time. Others did not have power and were not seen as important. Many of their stories are lost because they were not recorded, or written down. About 3,200 BCE, people started recording history with a system of writing called cuneiform. We know this because **archaeologists** have discovered ancient objects marked with this ancient writing.

People first started writing history on rectangular clay tablets using cuneiform.

Rewriting History

Trota was a woman who probably lived sometime in the 1000s or 1100s CE, in Salerno, Italy. She was a doctor who studied at the medical school in Salerno and wrote about treating diseases in women. Trota probably wrote one of the three books that became known as one book, *The Trotula*. Then, a strange thing happened. Over time, all copies of *The Trotula* had the names of male authors. It was not until the 1980s that historians started to wonder who really wrote the books. When they found out it was Trota, they helped take back her place in history.

*The Medical School in Salerno, Italy, was one of the few schools in the early **medieval** times to train both women and men.*

Keep Out!

In many cultures throughout history, science was off limits for women. Women were not allowed to study science or practice it. To get around these rules, some women worked in disguise. In the early 1800s, British women were not allowed to study medicine or join the army, but Margaret Ann Bulkley did both. Margaret was born in Ireland around 1790. When she was 19 years old, Margaret came up with a plan to study medicine. She changed her name to John Barry, after her uncle, and dressed as a man. The disguise worked. "John Barry" became a doctor and then joined the British army. John Barry worked as a surgeon and traveled to both South Africa and Canada. Dr. Barry died in 1865. A servant laid out the body for burial and discovered that John Barry was really a woman!

Margaret Ann Bulkley (left) disguised herself as a man in order to become a doctor. She called herself John Barry and became a great surgeon.

Assistants Only

To work in science, some women worked as assistants to men. Women were kept out of science clubs and **societies**. The French Royal Academy of Science did not allow women to be full members until as late as 1979. Even when Marie Curie won a Nobel Prize in 1911, she was still not allowed to be a member! More than 50 years later, Maria Goeppert Mayer became the second woman to win the prize for physics. It took another 55 years for Donna Strickland to become the third woman to win.

This book tells the stories of just a few of the amazing women scientists who have helped shape the world over the centuries.

CHAPTER 2

EPIC IN EUROPE

European women have been rocking the science world since ancient times. They had to fight for their rights to study and to work, and they didn't always get the credit they deserved. Yet, these women scientists are behind some of the world's most amazing discoveries. Let's get to know some of them.

Lise Meitner, Nuclear Powerhouse

Lise Meitner was born in 1878 in Vienna, Austria. From an early age, Lise loved math and science. She even kept a math book under her pillow! In the 1800s, girls were not allowed to go to school in Vienna after they turned 14. Lise had no choice but to stay at home and help look after her younger brother and to tutor, or teach, some younger girls.

Getting to University

By the time Lise was 21, Austrian universities accepted women students. To pass the entrance exam, Lise had to study all the subjects that boys at school had been taught, but that she had missed because she had had to stay at home. Two years later, after passing the entrance exam, she started her studies at the University of Vienna. In February 1906, Lise became the second woman to receive a **PhD** in physics from the university.

Lise never married, and she lived in Berlin for more than 30 years.

The University of Vienna was founded, or set up, in 1365. It took more than 500 years for the university to allow women to study there.

Now What?

Lise had earned her degree, but this did not help her find work—people were not happy about hiring a woman scientist. So, Lise turned to teaching again. She spent her evenings reading and studying physics problems. Then in 1907, Lise met the physicist Max Planck. A physicist is a person who studies physics. He suggested that Lise could work and study in Berlin, Germany.

Big City Move

Moving to Berlin was a big step for a shy, unmarried woman. Most women in the 1800s lived with their parents until they married. But moving to Berlin was life changing for Lise. Berlin was an amazing place for scientists in the early 1900s, and although Lise was still looked down upon because she was a woman scientist, she had many opportunities she would not have had outside of Berlin. Max Planck knew Lise had a brilliant mind, so he included her in the weekly meetings at his house. There, Lise and other scientists, including Albert Einstein, met to listen to music and talk about science.

HIDDEN HISTORY

A Secret Genius

If Lise had been a man, she would have been very famous in her time. But in the 1900s, even brilliant women scientists were not put in charge of work or taken seriously. In fact, when Lise once gave a scientific talk in Berlin, reporters jokingly wrote that it was about "Cosmetic Physics" instead of cosmic physics, or the physics of the universe. So Lise's great work remained hidden for much of her career. It is only in recent years that she has received the respect she deserved long ago.

A Woman's Workspace

Lise attended Max Planck's physics classes and also began working with the chemist Otto Hahn. As a woman, Lise was not allowed to work with Hahn in the laboratories at the Chemistry Institute. Instead, for two years, she worked in a small carpentry room that Otto helped her set up as a lab. Lise was not even allowed to go through the Institute's main entrance to reach the lab. Instead, she used a back door. And if she wanted to use the bathroom, she had to go to a hotel down the street!

Partners, but Not Equal

Lise worked for no pay until she became Max Planck's teaching assistant in 1912. That year, both Lise and Otto also started receiving salaries from the Institute for their research work. Although they were working partners, Otto's salary was almost three times more than Lise's salary. By 1917, Lise's salary finally matched Otto's. By then, she was also head of her own physics department.

WARTIME WORK

During the first year of World War I (1914 to 1918), Lise left Berlin for the battlefields in Austria. She worked as an X-ray technician, taking X-ray images of the injuries of wounded soldiers. Lise returned to Berlin in 1916.

Lise volunteered to work near the battlefields nursing the wounded during World War I. She would have worked in a ***military hospital*** *like this one.*

Race Against Time

During the early 1930s, things began to heat up in the physics world. Lise and Otto were working on a way to create new, heavy **elements**. In Italy, Enrico Fermi was trying to do the same thing. Scientists in England and France joined the race. Around the same time, Adolf Hitler and his **Nazi** party came to power in Germany.

Many male physicists who worked with Lise had great respect for her. The famous Danish physicist, Niels Bohr, helped Lise escape Berlin.

Running Away

In 1938, Lise's work was cut short as life for German Jews became very difficult. Even though Lise had become a Christian a few years earlier, she came from a Jewish family. This made her a target for the Nazis, who claimed that Jews were both troublemakers and not as important as non-Jewish Germans. At the same time, the Nazis banned scientists from leaving Germany. Lise knew she had to get out of the country and she escaped with the help of Niels Bohr. She hurriedly traveled to the Netherlands, carrying hardly any belongings in her rush to leave. A few months later, she arrived in Sweden. Soon after, World War II (1939 to 1945) broke out.

Lise and Otto worked together for more than 30 years.

Secret Letters

In Sweden, Lise took a low-paying job at the Nobel Research Institute of Physics. Meanwhile, she wanted to keep up with Otto's work in Berlin. She and Otto wrote to each other in secret, so the Nazis would not discover that Otto was in touch with her. Lise encouraged Otto to continue the experiments to create new elements. After getting some surprising results, Otto wrote to Lise for help.

An Amazing Discovery

Lise took the information and talked about it with her nephew, Otto Frisch, who was also a physicist. Lise carefully looked at the information in Otto Hahn's letter. She realized that nuclear fission had taken place in his experiments. In this process, the nucleus, or center, of an **atom** splits and releases energy. In February 1939, Lise explained nuclear fission in a paper she wrote for *Nature*, a leading science journal. Otto also wrote papers in Germany that supported Lise's work, but he was careful not to include Lise's name in the papers to protect her from the Nazis.

Going Nuclear

It did not take long for the word of the great discovery to spread. Scientists realized that Lise had unlocked the mystery of nuclear power. In August 1939, Albert Einstein wrote to U.S. President Roosevelt to warn him about the dangers of nuclear energy.

Otto and Lise wrote to each other often. Otto depended on Lise to help him figure out his experiments. Otto spoke to no one else about his work—Lise continued to be his science partner, even if from a distance.

The first atomic bomb was dropped on Hiroshima, in Japan, in 1945. Another was dropped on Nagasaki, Japan, shown here, shortly after.

Einstein was worried that the Germans would make an atomic bomb to use in the war. In the end, the Americans beat the Germans to it.

Missing Out on a Prize

During Lise's career, she was put forward for the Nobel Prize 48 times. Otto Hahn, Niels Bohr, and others put her forward over the years, but Lise never won. In 1944, Otto Hahn won the Nobel Prize for Chemistry for "his" discovery of nuclear fission. Tragically, because Otto had not mentioned Lise's name when he wrote his papers about nuclear fission, it was never clear that his work was mainly due to Lise. We will never know if Lise would have won the prize if she had been able to work in the open, rather than hiding away because of her Jewish background. However, we can make sure that today, Lise Meitner takes her rightful place as one of the most important and influential scientists of her age.

FAMOUS AT LAST

The science **community** finally recognized Lise Meitner's achievements in 1966. She, Otto Hahn, and Fritz Strassmann all received the Enrico Fermi Prize from the U.S. Atomic Energy Commission.

Today, there are awards given in Lise's name. The European Physical Society awards the Lise Meitner Prize every other year.

Rosalind Franklin, DNA Dynamite

Did you know that every person is a one-of-a-kind? This is because each person's DNA is different. DNA is a **molecule** in the **cells** of all living things. It contains all the instructions that cells need to live, grow, and **reproduce**. People knew about DNA back in the mid-1800s, but it was not until 100 years later that a scientist figured out what it looked like. The super-cool chemist behind that discovery was Rosalind Franklin.

An English Schoolgirl

Rosalind was born in London, England, in 1920. She attended one of the few schools that allowed girls to take science classes. By the time she was 15, she knew she wanted to be a scientist. Rosalind decided to attend Newnham College, a woman's college at Cambridge University. She graduated, or completed her course, in 1941.

Rosalind Franklin loved to explore outside the lab, too. She spent her spare time hiking, especially in France.

The European Space Agency (ESA) named their latest Mars rover Rosalind Franklin. Rosalind explored life on Earth through her DNA research. The rover is a robot that will look for signs of life on Mars.

Coal Is Cool!

Coal may not seem very interesting, but Rosalind's first job was to study it. In the early 1940s, Britain was busy fighting in World War II, and coal was needed as fuel for steam engines and ships carrying supplies to soldiers. The British wanted to discover what else coal might be used for. They asked Rosalind to study it. Coal was already used in gas masks, which had charcoal **filters**. These masks helped protect people from poisonous gas attacks in wartime. Rosalind's research helped improve them. She explained why some kinds of coal were better at keeping out gas than others. She published several papers about her findings.

On to Paris

After the war, Rosalind worked in Paris, France, for a few years. She loved her research there. In Paris, Rosalind learned about X-ray crystallography. This technique reveals how atoms are arranged in a **crystal**. This work would lead to the most exciting discovery of Rosalind's career.

BRILLIANT BUT BANNED

Cambridge University was founded in the 1200s, but women were not allowed to study there until centuries later. In 1869, the first Cambridge college to allow women opened, Girton College. However, the college building was far away from the main part of the university, where the men studied. It was clear that the university wanted to keep the men and women separate. And when they passed their exams, women only received the "title" of a degree and not a degree that was a full qualification. Full degrees were not given to women until 1948. The university even banned women from graduation ceremonies!

Going Home to London

Rosalind was sad to leave Paris. She loved the work, and she made good friends. Also, she never felt that being a woman scientist in France was a problem. Rosalind made the difficult decision to return to London. Her family was overjoyed to have her back. Rosalind had also heard about some interesting things going on in British research labs. In January 1951, she started working as a research assistant at King's College, London.

Men Only

Right from the start, Rosalind found working in London very different from Paris. King's College was like a men's-only club. The men who worked there had their lunch in the "common room." This was supposed to be a space for everyone, but women were not allowed in! Rosalind gritted her teeth and focused on her work.

Who's the Boss?

The director of the King's College lab, John Randall, asked Rosalind to work on a research project about DNA. Rosalind was an expert in the science of crystallography and Randall wanted her to apply this to studying DNA.

Rosalind used a microscope in her lab. Microscopes are tools that magnify things, or make them appear larger.

Rosalind thought she would be the lead person on the project, working with just one assistant. However, another scientist named Maurice Wilkins was also working on the DNA project. Maurice thought that Rosalind would be joining his existing team, rather than taking the role of lead scientist. Rosalind had no interest in working with Maurice, and continued her research with her assistant, Ray Gosling. Meanwhile, Maurice continued on his own work.

WHAT'S IN A NAME?

It is no secret that many of the men Rosalind worked with had a poor attitude toward women. This made her job difficult. But she never knew quite how disrespectful the men she worked with could be. Ten years after Rosalind died, the famous DNA scientist James Watson published his book about the DNA discovery. In it he gave Rosalind little credit. Instead, he referred to her as "Rosy," a name he never called her in person. In the book, he also said that he often wondered why she did not dress more stylishly and wear makeup!

Maurice Wilkins went on to become head of the Biophysics Department at King's College, shown here.

Life-changing Photo

In 1951, Rosalind gave a science talk in which she spoke about her findings on the structure of DNA, which showed that it had two strands that coiled around one another. In May 1952, Rosalind and Ray took an X-ray image of DNA and labeled it "Photo 51." The photo was proof of Rosalind's findings about the double-strand structure of DNA. At the same time, two British scientists working on DNA, James Watson and Francis Crick, suspected this structure. They had been carrying out their own research at Cambridge University.

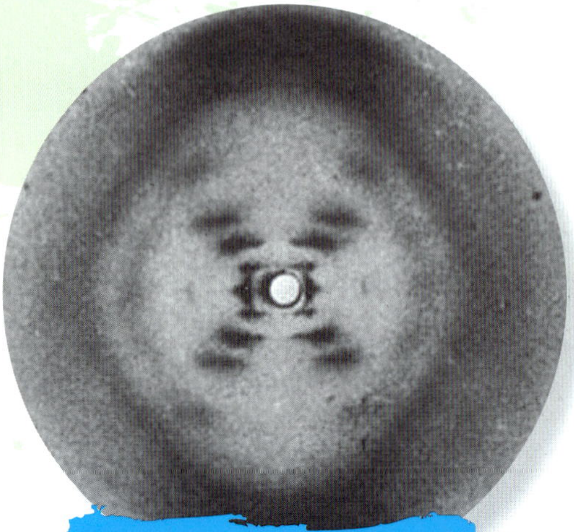

Rosalind's Photo 51 may not appear exciting to people who are not scientists. However, when James Watson saw it, he was thrilled. He knew that the photo gave him the proof he needed.

Proving an Idea

Somehow, Watson and Crick saw Photo 51. Rosalind did not know that her photo had fallen into the hands of the two scientists. The photo and Rosalind's paper proved that Watson and Crick's ideas about DNA were right.

The structure of DNA looks a little like a twisted ladder.

Then, in 1953, James Watson, Francis Crick, Maurice Wilkins, and Rosalind all published papers on DNA structure. In March 1953, Rosalind and Maurice visited James and Francis to see the DNA model that the Cambridge scientists had made.

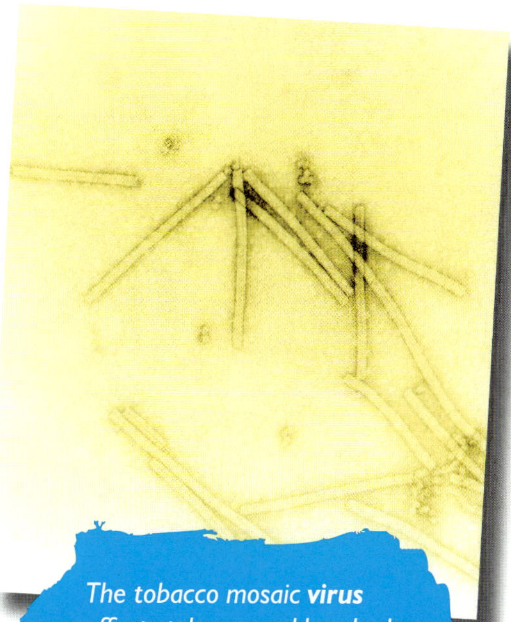

*The tobacco mosaic **virus** affects tobacco and hundreds of other plants. The name "mosaic" comes from the green and yellow pattern that the virus causes on leaves.*

Life After King's College

Not long after all of this excitement, Rosalind left King's College. She had been unhappy there from the beginning, and she was relieved to get out. Her new job at Birkbeck College in London offered her a chance to dive into new research. Rosalind stayed in touch with James Watson and became friends with Francis Crick and his wife.

A Tragic End

For the next few years, Rosalind researched and published many papers. She traveled to the United States twice to present her papers and meet fellow scientists. Then, in the fall of 1956, Rosalind received some devastating news—she had cancer. Rosalind died in April 1958 at just 37 years of age. Four years later, Watson, Crick, and Wilkins received the Nobel Prize for their discovery of the structure of DNA.

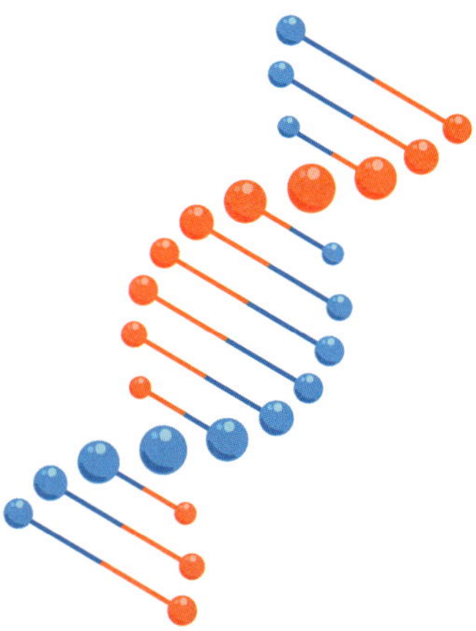

HIDDEN HISTORY

Not Just DNA

Rosalind's research after King's College is rarely talked about. Yet, she continued to carry out important work on viruses. By looking at the tobacco mosaic virus, she hoped to find out more about other viruses. Before she died, Rosalind was working on the structure of the polio virus. Polio is a devastating disease that can cause **deformity** of the limbs.

Maria Merian, Insect Detective

Maria Merian grew up drawing. She was born in 1647 in Germany into a family of artists. Her father died when Maria was young. Her mother then married another artist. Maria's stepfather taught her how to paint, and Maria became an expert. What made Maria different from other artists was her interest in insects. Her first drawings were of the silkworms that fed on the mulberry bushes in her garden.

Maria's book about the insects of Suriname became a bestseller. This is an example of one of her paintings.

Maria was also interested in caterpillars. She watched how they metamorphized, or changed, into butterflies. In the 1600s, scientists did not understand how this happened. They believed in something called "spontaneous generation." This meant they thought that insects grew, fully formed, out of mud or water.

Taking a Close-up Look

Maria realized that insects were very special. She took her time studying them, and she drew every detail. In this way, Maria became one of the world's first entomologists, or people who study insects.

Maria drew her illustrations and then painted them with watercolor paints.

During her lifetime, Maria drew and described more than 200 **species** of insects and amphibians, which are creatures that can live in water and on land. By watching and drawing these creatures, she was able to explain how they lived, why they lived there, and why they behaved as they did.

Drawing from Life

Maria stood out from other scientists of her age who were interested in the natural world, because she drew living insects in the places in which they lived. Everyone else at the time studied dead insects that they had collected. Maria's drawings showed the relationship between plants, insects, and animals for the first time.

AN INDEPENDENT WOMAN

Maria was very independent for a woman in the 1600s. She left her husband in Germany and moved to Amsterdam, the Netherlands, with her daughters. There she was able to sell her insect drawings. After years of working and saving her money, Maria and one of her daughters traveled to Suriname, in South America. Maria stayed there for two years, studying and drawing the plants and animals she found there, especially insects. Many of these South American species were not known in Europe in the 1600s.

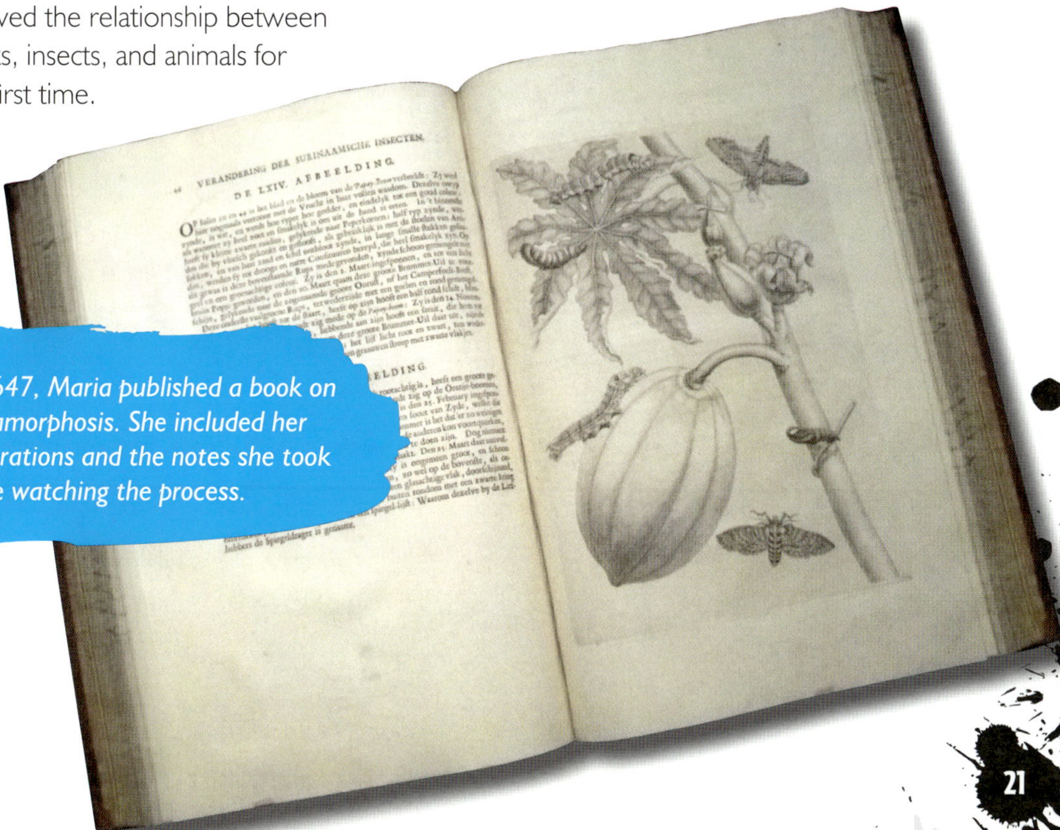

In 1647, Maria published a book on metamorphosis. She included her illustrations and the notes she took while watching the process.

Inge Lehmann, Mover and Shaker

Inge Lehmann was born in 1888, in Copenhagen, Denmark. A few years before she was born, scientists had developed a machine that would help shape Inge's future—the seismograph. This device records **vibrations** beneath Earth's surface, such as earthquakes and the **seismic waves** they produce. Although earthquakes are rare in Denmark, Inge had felt a small one when she was a teenager. She became fascinated by them, and about 20 years later, became Denmark's first seismologist, or a person who studies earthquakes.

JUST LIKE THE BOYS

Inge's school days were spent at a mixed-sex school. Both boys and girls attended the school, and Inge learned math and science, alongside the boys. Inge thought this was what all girls did, and it was not until much later that she discovered it was unusual. The teachers at the school encouraged Inge to keep studying science. She went on to study math, astronomy, physics, and chemistry at the University of Denmark.

Computers were not yet invented when Inge was working. She put all her seismic information on cards and organized the cards in empty cardboard boxes. Inge then carried out all her calculations using the information on the cards.

Seismic Sensation

Inge worked as a professor's assistant before getting her first real work in seismology. In 1925, she got a job setting up Demark's first seismic stations. She set them up near Copenhagen and prepared the equipment to be set up in Greenland. At the same time, Inge studied seismology. When she graduated, she became head of the Seismological Department at the university.

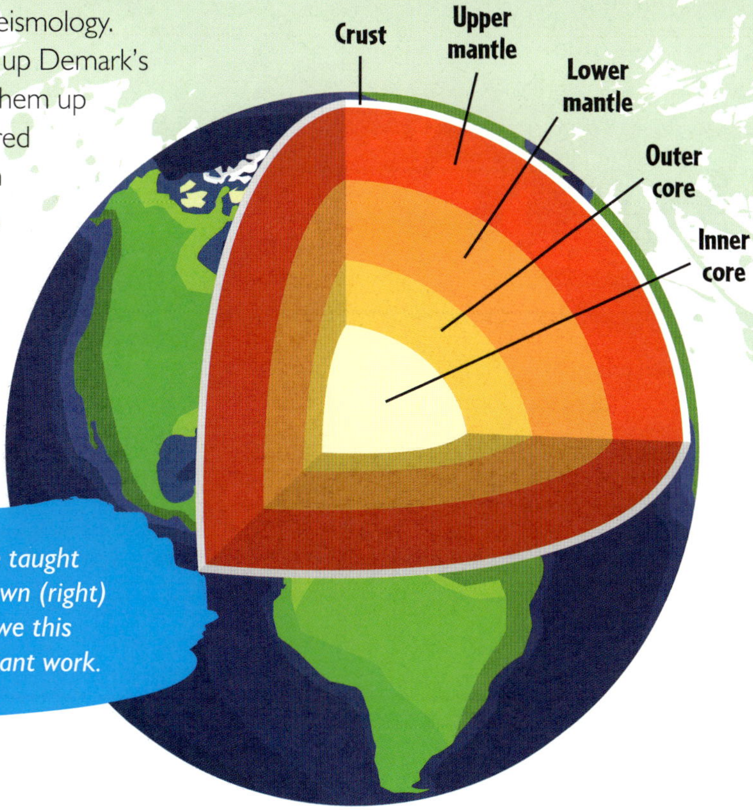

Today, schoolchildren are taught about Earth's layers, shown (right) in this illustration. We owe this knowledge to Inge's brilliant work.

A Solid Discovery

In her new role, Inge started collecting information about seismology. Part of this was figuring out how long it took for seismic waves to travel through Earth. In 1929, when a major earthquake shook New Zealand, Inge noticed something weird about it. Some of the waves traveling away from the earthquake were bent or traveling at a strange angle. Over the next few years, Inge studied the information she had gathered and concluded: the inside of Earth is not all hot liquid. It also has a solid, metal core, or center, surrounded by liquid.

Unstoppable Inge

Inge kept working on seismology long after she retired, or finished work. Using new, improved equipment, Inge made another discovery. Earth has a layer that sits about 120 miles (200 km) below the surface. This is called the Lehmann discontinuity.

CHAPTER 3

AWESOME IN ASIA AND AUSTRALIA

You may know that Asia and Australia have a long history of scientific brilliance. But did you know that for centuries many awesome women there have been hard at work putting their mark on the scientific world? Let's meet some of these superstars of science.

Wang Zhenyi, Star-gazing Poet

In 1768, a young Chinese astronomer was born. And she was one who refused to let old-fashioned ideas get in her way. Wang Zhenyi used brains and determination to become one of the greatest minds of the time. In the eighteenth century, education in China was for rich boys only. Luckily for Zhenyi, the Wang family was different.

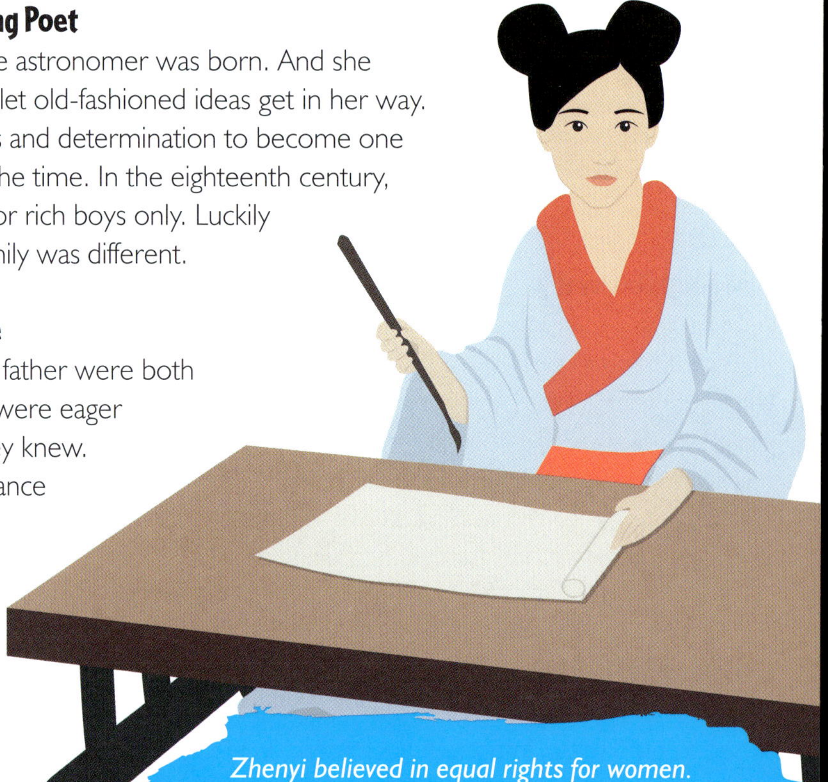

Home-schooled Heroine

Zhenyi's grandfather and father were both well educated, and they were eager to teach Zhenyi what they knew. Zhenyi jumped at the chance to learn from them. She studied astronomy with her grandfather and math with her father. She taught herself, too, and read all the books she could find, especially ones about math and astronomy.

Zhenyi believed in equal rights for women. In one of her poems, she wrote, "It's made to believe, women are the same as men; Are you not convinced, daughters can also be heroic?"

In Zhenyi's time, most people in China thought that angry gods caused eclipses. Zhenyi proved that eclipses happen because of the movement of the Moon around Earth.

Solving Moon Mysteries

Zhenyi thought **eclipses** were amazing, but the books she read could not explain them. So she did what any scientist would do—she did her own investigation! Zhenyi already had ideas about how eclipses worked, but she wanted to prove them. To do so, she made a small model of an eclipse. She used a lamp for the Sun and hung it above a round table, which represented Earth. Between the lamp and table, she used a mirror for the Moon. By moving the mirror, Zhenyi explained how eclipses work.

NOT JUST A BOOKWORM!

Zhenyi did not spend all her time with her nose in a book! She also liked to be active. Zhenyi learned **martial arts** from the wife of an army general. She also learned how to shoot a bow and arrow and ride a horse.

Fighting for Equality

Zhenyi's father took his daughter with him as he traveled around China. On their journey, she saw how the poor, especially women, lived in the countryside. She decided to write poetry to **protest** about these living conditions. Important people in China were not very happy about Zhenyi's poems when they discovered them, but Zhenyi did not care. She used her powerful writing to tell people how hard life for the poor and women was in China. When Zhenyi died at age 29, she had written enough poems to fill 13 books. Many of these poems survive today.

Ruby Payne-Scott, Making Waves in Australia

Few people in the world took **radio astronomy** seriously until the 1940s. Australia's Ruby Payne-Scott was one of the first scientists—and the only woman—to realize that this new science was the future of astronomy.

Mad About Math

Math was always Ruby's best subject in school. At her all-girls' high school in Sydney, Australia, Ruby was top of her class. Before she was 16, Ruby received a scholarship, or a prize that pays for a person to study, to attend the University of Sydney. She graduated in 1933 with an honors degree in physics. She became only the third woman to receive the degree from that university.

Despite the fact that she was a woman working with many men, Ruby led the research in a lot of the places in which she worked.

Ruby was the first woman ever allowed to carry out research at the Australian Wireless Company.

Ruby carried out radio astronomy experiments on the cliffs of Dover Heights in Sydney, Australia.

Back to School

Next, Ruby got a job. She worked as a part-time research assistant at the Cancer Research Committee (CRC). Just a few years later, the CRC closed down, and Ruby had to find another job. It was 1936, and no one wanted to hire a woman physicist. Ruby had to change direction, and went back to university to earn a teaching degree.

Wartime Working

Ruby taught only for a year before she left to work for a wireless company. It was not her dream job—she had to work as a librarian at first—but it was a way back into scientific research. About two years later, a much better opportunity came up. Many thousands of Australian men were away fighting in World War II, which meant that women stepped in to do their jobs at home. The war also created the need for research into new weapons. In 1941, Ruby started top-secret work for the Council for Scientific and Industrial Research (CSIRO). Her research helped develop better **radar** to track enemy fighter planes.

FIGHTING FOR RIGHTS

During the war, the CSIRO paid women scientists the same salary as the men. But when the war ended, the CSIRO went back to paying the women less. Ruby protested. The CSIRO finally agreed to pay the women the same amount as the men, but only if they stayed in the job they had worked during wartime. Ruby tried to get the same rights for new women employees, but the CSIRO refused.

Focus on Space

During the last year of World War II, Ruby turned her attention to astronomy. In March 1944, she and another scientist, Joseph Pawsey, set up a small experiment to see if they could detect any radio waves from space. The test did not lead to any amazing findings, but it was a start. A year later, Ruby and her team moved outside the lab. This time, they were better prepared, and they had better equipment.

Women's Wear

By now, Ruby was a top-notch physicist doing awesome work. Still, she faced a lot of **discrimination** because she was a woman. Her male bosses even tried to tell her what to wear! During the hot, Australian summers, the male scientists wore shorts. The CSIRO rules stated that women had to wear dresses or skirts at all times. Ruby ignored them. She made a point of wearing shorts, which were much more practical.

This photograph shows Ruby at a physics meeting in 1952. What do you notice about this picture?

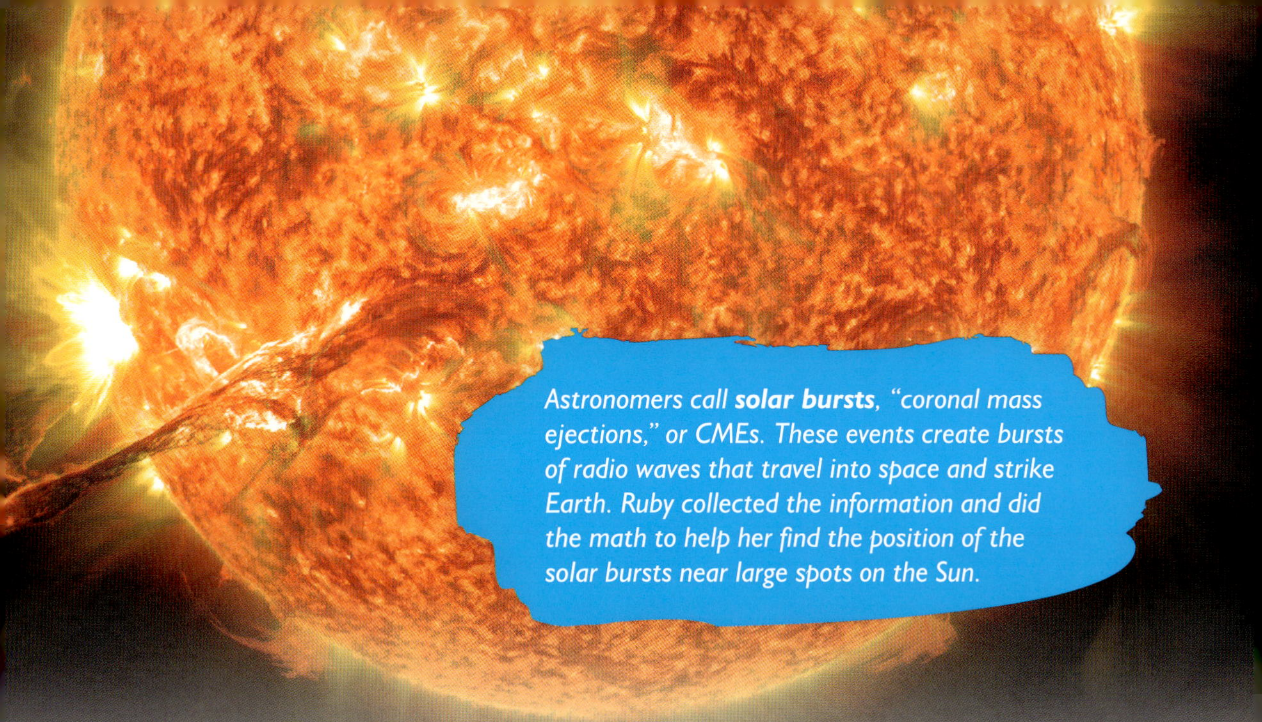

*Astronomers call **solar bursts**, "coronal mass ejections," or CMEs. These events create bursts of radio waves that travel into space and strike Earth. Ruby collected the information and did the math to help her find the position of the solar bursts near large spots on the Sun.*

Bursting with Energy

Ruby's most amazing discoveries came within the next few years. At Dover Heights, she and her colleagues built and set up their equipment on the side of the cliff. They used a device called an interferometer. This picked up high-energy bursts coming from the Sun. These waves can be detected only with radio astronomy. Ruby helped to design another, better interferometer. This one searched the sky many times per second, and Ruby discovered two more types of solar bursts.

Ruby Who?

Ruby's work was groundbreaking, yet most people do not know her name. This may be because she left her job to start a family, just as radio astronomy was taking off. Meanwhile, men like her coworker, Joseph Pawsey, became famous. Ruby never returned to research. Instead, she raised her two children and went back to teaching in a school—and no one there knew anything about her amazing former career. Ruby died in 1982. In 2013, the Australian Institute of Physics introduced the Ruby Payne-Scott award to recognize the work of other brilliant young physics researchers.

HIDDEN HISTORY

No Work for the Mrs.

When Ruby married in 1944, she had to keep it a secret. Women were expected to **resign** from government work as soon as they married. Ruby kept her ring on a chain around her neck and hid her marriage for two years so she could continue to work. She was forced to leave for good when she became pregnant with her first child.

CHAPTER 4

AMAZING IN AMERICA

For hundreds of years, science was a men-only club. To study and work in the field, women had to overcome unbelievable obstacles. Let's meet some of the determined women scientists who broke down barriers and changed the rules in North America!

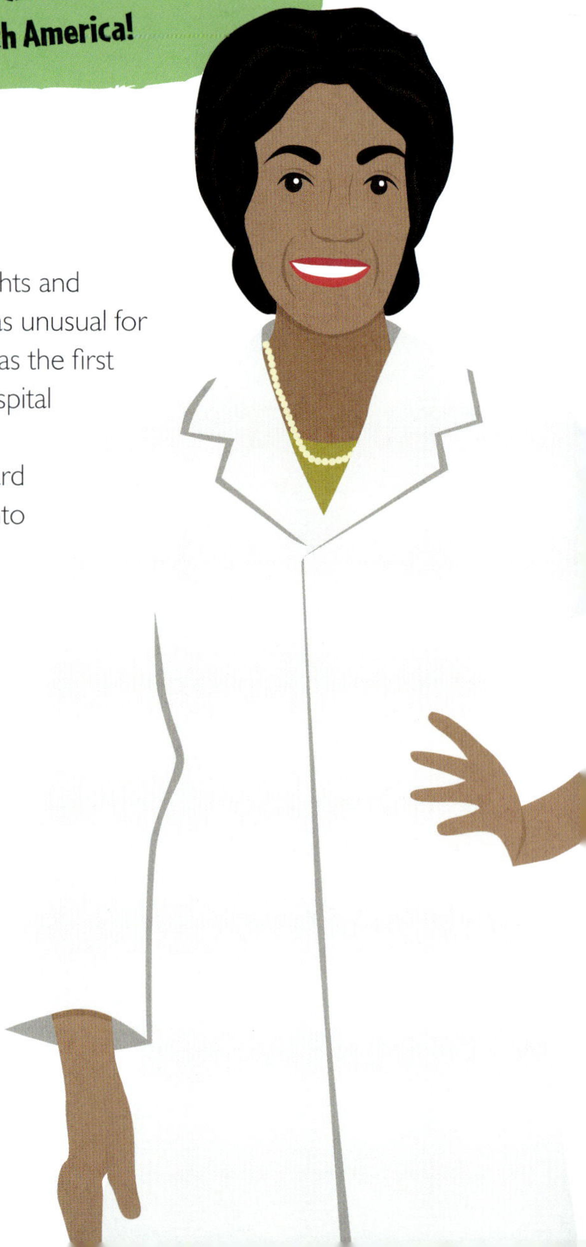

Jane Cooke Wright, Cancer Crusader

When Jane Cooke Wright was born in 1919, African Americans were still fighting for basic rights and **segregation** was widespread. Jane's family was unusual for the time. Her father, Louis Tompkins Wright, was the first African American doctor to work at a public hospital in New York City. He was also one of the first African Americans to earn a degree from Harvard Medical School. Louis's father had been born into slavery, but he, too, earned a medical degree.

Thinking Outside of the Box

Jane was a brilliant student and won a full scholarship to the New York Medical School. After graduating, she started her cancer research. Jane joined her father at first, but soon began her own experiments. She was one of the first people to realize that not all patients, or cancers, were alike. Her realizations led to new treatments.

*Jane became known as the "mother of **chemotherapy**" for her work in cancer treatment.*

This is a sculpture of Jane's father, Dr. Louis Tompkins Wright. Jane started out studying for an art degree, but her father encouraged her to become a doctor. Jane switched into medicine and became one of the school's top students.

LEADING THE WAY

Not only did Jane make breakthroughs in the lab, but she also took on leading roles. When she was just 33 years old, Jane took over the Cancer Research Foundation that her father started. She became head of the Cancer Chemotherapy Department and the Associate Dean of New York Medical College. Jane was also the first woman president of the New York Cancer Society and one of the people who began the American Society of Clinical Oncology (ASCO).

Early Days of Chemo

Mustard gas was a weapon used during both World War I and World War II. It caused severe burning to the skin, eyes, and lungs. However, scientists discovered something surprising about the drug during the war. **Chemical compounds** related to mustard gas seemed to work against certain types of cancer. This breakthrough led to a new cancer treatment called chemotherapy. This is where Jane came in. Her research helped get chemotherapy from the lab into the hospitals, where it could be used to treat people. Today, chemotherapy is one of the leading treatments for cancer patients.

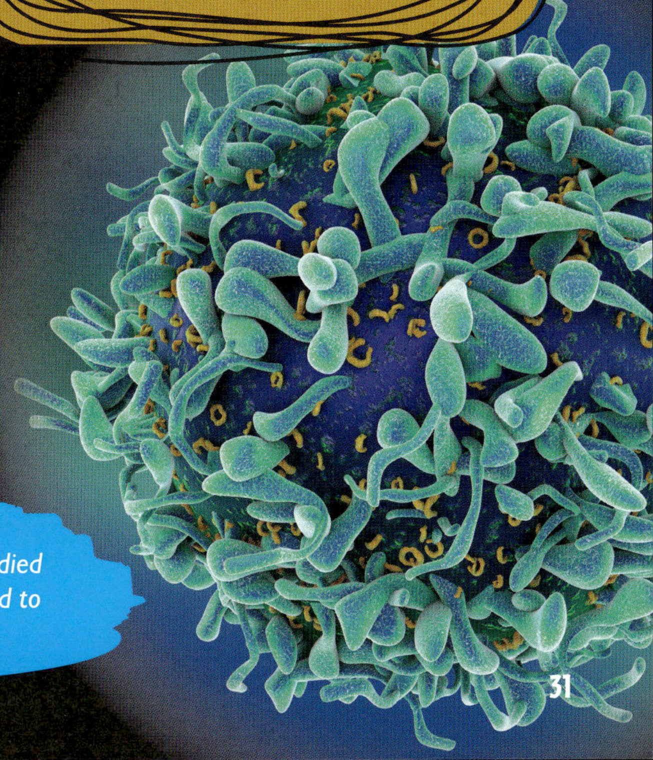

Cancer cells are not all alike. Jane studied them to see how different cells reacted to different treatments.

Ruby Hirose, Research Rockstar

When Ruby Sakae Hirose was a young girl, polio was one of the most feared diseases. Polio spreads easily and has caused widespread **epidemics**. In 1955, Jonas Salk became famous for introducing the polio **vaccine**. Much less famous is Ruby Hirose, yet her research helped pave the way for this lifesaving vaccine.

During World War II, the military rounded up Japanese-American families, mainly in the western United States, and took them to camps. Ruby's family was among the Japanese taken to these camps. Ruby was living in Cincinnati, Ohio, at the time, where she was carrying out research. She was not sent to a camp.

From Japan

Ruby's family came from Japan to the United States in 1904. Ruby was born later that year, near Seattle, Washington. New Japanese **immigrants** like Ruby's parents were called issei. Their children, who were born in the United States, were called nisei. Along with many other issei, the Hirose family became farmers. As a Japanese immigrant, Ruby's father was not allowed to own or **lease** his own land.

Hardworking and determined, Ruby devoted her life to finding cures for difficult and dangerous conditions.

Moving East

Ruby's father encouraged her to go to university, but he had very little money to give her. Ruby worked part-time while she studied at the University of Washington. She aced her classes and knew she wanted to keep studying science. She stayed in Washington to do her master's degree, but then moved to Ohio to complete her PhD.

Lifesaving Vaccine

Ruby started working and teaching at the University of Cincinnati. She soon made a name for herself working on various medicines. Her research on a polio vaccine was one of her most important projects. Other scientists started talking about the quiet research scientist who was doing such brilliant work. In 1940, the American Chemical Society recognized Ruby for her contribution to chemistry.

Ruby studied how Indigenous peoples used the goldenseal plant as medicine. She wrote a paper on how the plant helped heal skin diseases and sore eyes.

NO MORE ITCHY EYES

Ruby suffered from **allergies**. Her red eyes and runny nose made her determined to find some relief. She was already researching a vaccine for a disease called diphtheria when she came up with an idea to treat her allergies. She added a new chemical compound to an allergy medicine, which made it much more effective. The new medicine made all the difference for people with allergies.

Gertrude B. Elion, Disease Destroyer

Gertrude Elion had just finished high school in the Bronx, New York City. Like many high school students, she was wondering what to do with her life. Then her grandfather died of stomach cancer. Gertrude made her decision—she was going to learn how to fight, and maybe cure, diseases like cancer.

The only problem with Gertrude's plan was money. Her father had been a successful dentist. But the **Great Depression** struck the United States in 1929, and suddenly, her family had little money to spare. Gertrude found out that Hunter College in New York offered free **tuition**. She applied and graduated with a degree in chemistry when she was just 19.

During the Depression, people were out of work and so poor that they could not afford food. They lined up for meals at soup kitchens like this one.

By the time she retired, Gertrude had developed 45 different drugs.

Never Give Up!

Around the same time as she was celebrating her graduation, Gertrude became engaged to be married. Then tragedy struck. Gertrude's young fiancé died. Now she was more determined than ever to get involved in medicine. She also knew that she needed at least a master's degree to get a job in a chemistry lab. During the Depression, universities did not have a lot of money, so they only offered to help male students by partly paying for their studies.

No Stopping Gertrude

Gertrude did not let the many setbacks she came across stop her. She volunteered at a chemistry lab to get experience. She worked there and at other jobs until she had saved enough money to start on her master's degree. And when she started that degree, Gertrude worked part-time as a secretary and a teacher to earn money.

WOMEN IN WARTIME

During World War II, millions of men went to fight in the war, including many scientists. This opened the door for women scientists who may not have been hired otherwise. Gertrude was one of them. She worked as a chemist in two jobs before being hired in 1944 to research new medicine. She stayed in this job for the rest of her career.

Although she retired, Gertrude still taught chemistry students and helped at the lab whenever people asked her to. She also worked for the World Health Organization (WHO). Her research helped develop a drug to treat malaria. Malaria is a deadly disease carried by mosquitoes.

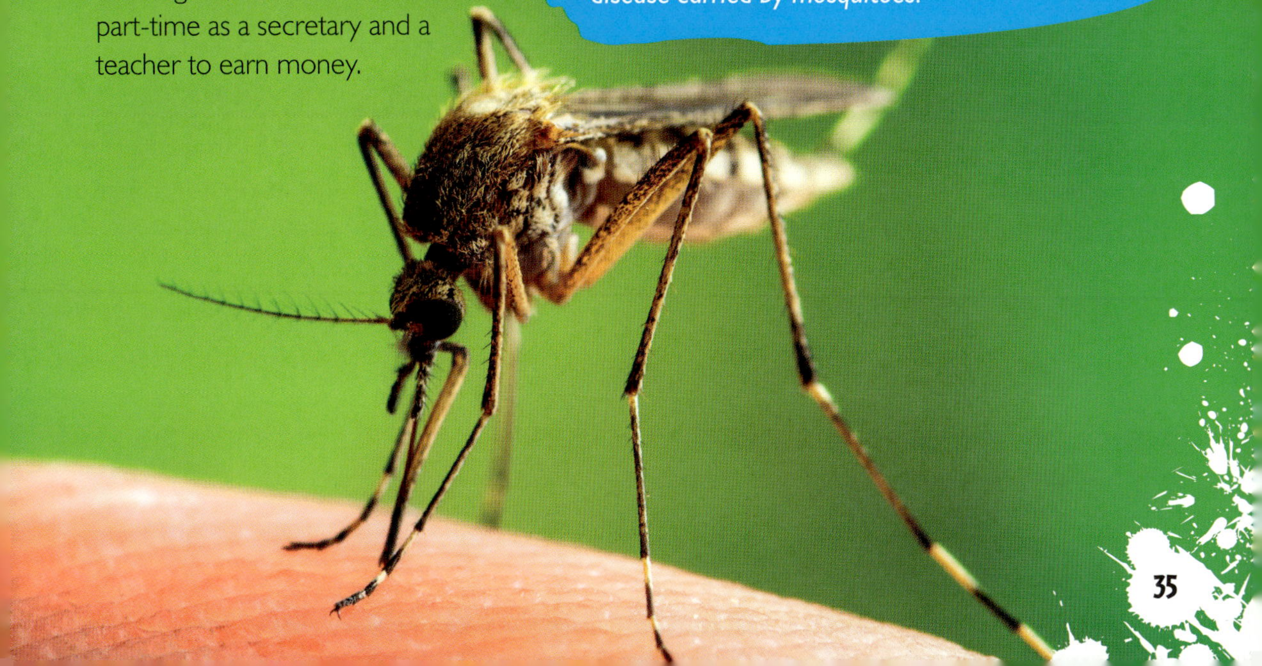

Making Medicine

Gertrude's first real chemistry job was testing food products. She was grateful for the work, but testing pickles was not her ideal job! In 1944, the chance of a lifetime came up. Gertrude landed a research job at a **pharmaceutical** company. She could finally start making medicine. Gertrude's lab partner was another chemist, George Hitchings. Together, they changed the way that people created medicine.

Killing Cancer Cells

What made Gertrude's research different? It concentrated on making drugs that destroyed only unhealthy cells in the body. This was cutting-edge thinking in the late 1940s. In 1950, Gertrude and George created two new drugs to treat leukemia, a form of cancer that affects mainly children. These drugs inspired other scientists to do even more research. Today, Gertrude's work has helped cure 80 percent of children with leukemia.

A Lab of Her Own

Together, Gertrude and George created drugs to treat many other diseases. They were a great team, but when George retired, Gertrude took over as the head of the department. She kept researching medicine for all kinds of diseases, but some of her most amazing work focused on treating viruses.

Gertrude and George figured out how unhealthy cells spread through the body. This enabled them to create drugs that kept the cells from spreading.

In 1988, Gertrude became one of only 48 women who have won a Nobel Prize for science.

Scientists did not think that drugs could help viral diseases. They thought the drugs would be too hard on the body. Gertrude proved them wrong and developed drugs for viral diseases such as chickenpox. Gertrude's drug research helped other scientists to develop life-changing treatments. This made Gertrude very happy.

Changing the World

At 15, she had dreamed of changing the world for the better. In her long career, she did just that. Almost 20 years after Gertrude began working in pharmaceuticals, she won the Nobel Prize for Chemistry. She shared the prize with James Whyte Black and her long-time research partner, George Hitchings.

BECOMING DR. ELION

Gertrude took classes toward her PhD, but the university made her choose between work and study. Gertrude loved her job as a research scientist, so she gave up on her dream of a receiving a PhD. This turned out to be a great decision. In 1969, both George Washington University and Brown University awarded Gertrude honorary doctorate degrees.

Marie Tharp, Deep Thinker

The ocean floor lies more than 12,000 feet (3,650 m) below the surface of vast stretches of salty water. Some of the deepest parts of the ocean are three times that deep! This dark, underwater floor is not an easy place to study. For years, people thought the bottom of the ocean was flat and muddy. Then along came Marie Tharp, the woman who changed all that.

Moving Around

Marie was born in 1920 in Michigan, but she and her family moved all around the United States. This was not easy for Marie because it meant she had to go to many different schools while growing up. But she loved going to work with her father. He made soil **survey** maps for the U.S. Department of Agriculture. In spite of this, Marie did not think about science when it came time for college. Her first degree was a double major in English and Music.

Marie used her math and drawing skills to turn pages of numbers into hand-drawn maps that showed the mountains and valleys of the ocean floor.

Maybe She's a Scientist After All

With her arts degree in hand, Marie thought about jobs she could do. None of the **traditional** women's jobs were appealing, but what about other jobs? Marie read something about working in the oil industry, but she needed a **geology** degree for that. So, she went back to school for a few years, first for a geology degree and then for a math degree.

Marie did get a job with an oil company, and she worked there for three years. It was not the exciting challenge she hoped it would be. Marie started looking for something more interesting. Columbia University in New York was hiring a geologist, or a person who studies geology, so Maria applied. She got the job and started working at the university's Lamont Geographical Observatory.

A WOMAN IN A MAN'S WORLD

Marie thought that her job at Columbia would put her in the middle of an exciting new project. Instead, she found herself doing the math for the male scientists to use in their research. Marie was a bright, experienced scientist by this time. Yet, the men still expected her to make coffee when she was not busy doing other jobs!

Marie's maps revealed a first, detailed picture of the ocean floor. However, even today, more than 80 percent of ocean floor remains unmapped.

Chance of a Lifetime

Marie's life at Columbia changed completely when another geologist, Bruce Heezen, came to see her. Bruce wanted Marie to help make a new map. But this was not just any map. He wanted to make a map of the ocean floor. Bruce had many thousands of measurements gathered by the U.S. Navy during World War II. The world's ocean floor is so vast that Marie needed thousands more measurements to make her map. Bruce continued to collect information and Marie worked on the map.

Mapping Marvel

Marie mapped the Mid-Atlantic Ridge, an underwater mountain range in the middle of the Atlantic Ocean. As she did so, she found something incredible. Running along the middle of the mountains, there was a deep valley. Marie later compared her map to another map of underwater earthquakes. She saw that the earthquakes lined up with the valley. This was the first real evidence of **continental drift**.

The Navy and Columbia's research team measured depths using sound waves. These measurements helped Marie know where the ocean floor rose into underwater mountains and fell into plains and valleys. Because she was a woman, Marie was not allowed on the research ships, like this one, until 1965.

You're Fired!

Marie was excited to share her findings, but her boss Maurice Ewing did not believe her. He and many other scientists at the time did not believe in continental drift. Bruce eventually published the map of the North Atlantic Ocean in 1957. Maurice Ewing was angry and took away the funding for Bruce's department. And he fired Marie.

Continental drift is caused by plate tectonics. About 100 million years ago, the plates forced Earth's land to break apart and form the continents. They still move today.

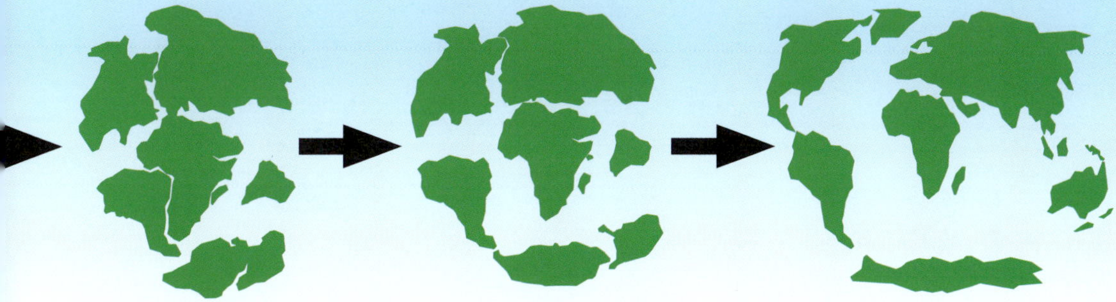

This did not keep Marie from her maps. She and Bruce continued to work together, but Marie worked at her house. In 1977, Marie and Bruce published a masterpiece: the World Ocean Floor map. The map showed the 40,000-mile (64,374 km) long ridge that runs around the world. Marie continued her mapping work, but her greatest work was done. In 1996, the Society of Women Geographers gave Marie an "Outstanding Achievement Award."

From her mapping work, Marie also collected important information about tectonic plates, which are large sections of Earth's crust that lie beneath its continents. Earth's tectonic plates are the white lines marked on this map.

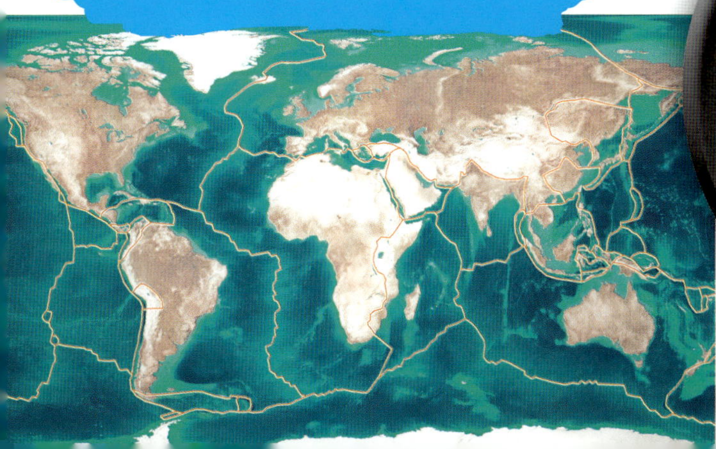

HIDDEN HISTORY

Just "Girl Talk"

Even Marie's friend and coworker, Bruce Heezen, argued with her about the idea of continental drift. He called her ideas "girl talk." The scientists who went on to prove this idea in the 1960s used Marie's maps. Marie's contribution was not recognized until 1997 when the Library of Congress recognized her with an award.

CHAPTER 5

ACING IT IN AFRICA

Did you know that women in ancient Egypt were educated in science, just like men? That was pretty forward thinking for the time! Africa has given the world some awesome women scientists. So, let's meet a couple of them!

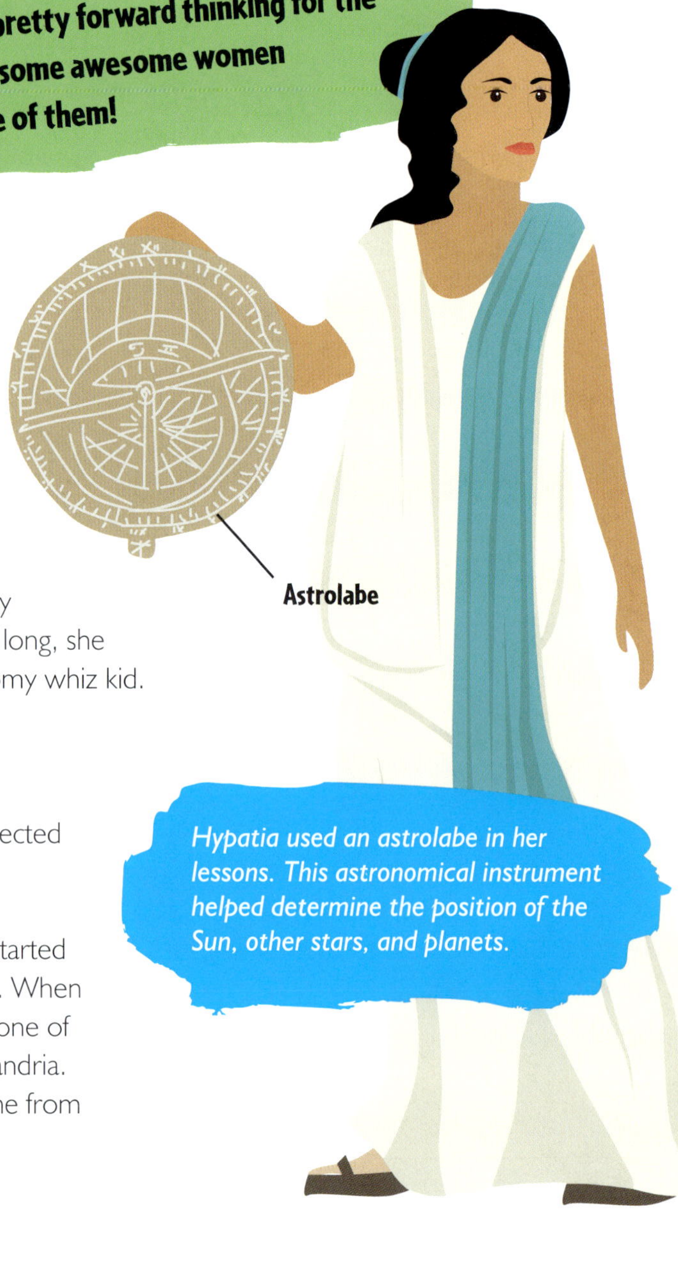

Hypatia, Greek Mathlete

Back in the 300s CE, Alexandria, Egypt, was the place to be. The city was the go-to for all the top scientific minds of the time. Hypatia, who was born sometime between 350 and 370 CE, was one of Alexandria's greatest scientists. She started learning about math and astronomy when she was very young. Before long, she was Alexandria's math and astronomy whiz kid.

Top Teacher

Hypatia learned from her father, Theon. Theon was a brilliant, respected astronomer, mathematician, and professor. He taught his daughter everything that he knew. Hypatia started helping her father with his writings. When she was old enough, she became one of the most popular teachers in Alexandria. The word spread, and people came from all over to learn from her.

Hypatia used an astrolabe in her lessons. This astronomical instrument helped determine the position of the Sun, other stars, and planets.

A Tragic End

Hypatia's brilliant life came to a sudden, violent end one day in 415 CE. She was traveling in her carriage (shown right) when a group of angry Christians surrounded the carriage. They dragged Hypatia out, pulled her into a church, and beat her to death. There are many ideas about why they killed Hypatia, but no one knows for sure.

After Hypatia's death, many historians wrote about her. Most agree that Hypatia was a great thinker and teacher. None of Hypatia's books have survived, but we know she wrote about math and astronomy. One of her works included astronomy tables that described the movements of objects in space. According to her student, Synesius, she also used or maybe invented astronomy tools, such as a planisphere.

A planisphere is a tool that shows the location of the stars in the night sky. The disks move to any date and time to show a map of the stars at that time.

FRIENDS AND ENEMIES

Hypatia had powerful friends. This put her in the middle of a power struggle in Alexandria. Hypatia knew and advised the governor of Alexandria, Orestes. Meanwhile, the Christian bishop, Cyril, wanted to control Alexandria. He forced the city's Jews to leave. He was also suspicious of the non-Christians, especially Hypatia who taught non-Christian ideas. Early Christians did not approve of this kind of study or of a woman teaching and spreading these ideas. These facts alone may have led to her death.

Wangari Maathai, Eco Warrior

Ihithe is a small village in Kenya, Africa. The country was still British-controlled when Wangari Maathai was born in Ihithe in 1940. Wangari's parents were farmers, like many others in the village. The land was rich. Years later, Wangari worked to protect the land in Kenya.

Most village girls did not go to school, but Wangari was different. Her family sent her to school when she was eight. She was a bright student, but there were few opportunities for girls in Kenya. However, Wangari won a scholarship to go to a university in the United States. Wangari graduated with a degree in biology. When she returned to Africa, she became the first woman in East and Central Africa to earn a PhD.

Wangari inspired many other groups around the world through her continuous hard work to protect the natural world.

Wangari won the Nobel Peace Prize in 2004. She was the first African woman to win the Prize.

Protecting the Planet

Wangari's degree was in **veterinary** sciences, and she started teaching at the university in Nairobi, Kenya. At the same time, Wangari was worried about **deforestation** in Kenya and in 1977, she set up the Green Belt Movement. The organization paid poor women in small communities to plant trees. The Movement spread to other African countries and, today, more than 51 million trees have been planted.

ARRESTED FOR CARING

Wangari's Green Belt Movement helps both women and the environment. By the late 1980s, it was growing quickly. The government became worried that women were becoming too independent. Wangari and other women protested against the government. The police arrested Wangari and put her in jail. She was released, but the government kept trying to shut down her Movement. However, Wangari and her supporters refused to let them, and the Movement lived on.

Getting into Government

Kenya's government did not support Wangari's **environmental** work. So she decided to make a change and get into government herself. In 2003, Wangari became assistant minister of the environment, **natural resources**, and wildlife department. She continued to fight for the environment and to work for women's rights until she died in 2011.

Wangari made a huge difference to the environment in Kenya.

GLOSSARY

allergies Conditions that make people sick when eating, touching, or breathing in certain substances
archaeologists People who study human history by looking at the places where they once lived and the objects they left behind
astronomy The study of objects in space, such as stars and planets
atom The smallest part of a chemical element
cells The basic units of which all living things are made
chemical compound Something that is made of two or more elements
chemistry The science that looks at what makes up substances and how they change when they are put together
chemotherapy A form of cancer treatment
comets Icy objects that travel through space
community A group of people who live or work together
continental drift The slow movement and separation of Earth's continents
crystal A solid material with molecules that fit together in a repeating pattern
cultures Groups of people with the same arts, language, and religious beliefs
deforestation When forests are destroyed and no new trees are planted
deformity When something does not have an expected shape or form
discrimination Treating a person or group unfairly, for example, because of skin color
eclipses Events in which the Sun appears dark because the Moon is between it and Earth; or, when the Moon becomes darker when Earth's shadow moves over it
elements Simple chemical substances that cannot be separated into smaller chemical parts
environmental Related to protecting the natural surroundings, such as air, water, and plants
epidemics The spread of particular diseases among a large group of people at the same time
exoplanets Planets or stars that are outside our Solar System
filters Devices that capture unwanted substances
genetics The study of how DNA and genes work and are passed down through generations
geology A science that studies Earth, particularly rocks
Great Depression A period in the 1930s in which many people did not have jobs and became very poor
immigrants People who come to a country from another country to live
laboratory A place with special equipment for scientific experiments
lease To be allowed to use a property for a period of time for a fee
martial arts Traditional forms of hand-to-hand fighting without the use of weapons
medieval A period of time between fifth and fifteenth centuries
military hospital A special hospital for wounded soldiers and others in the armed forces
molecule A group of atoms joined together
natural resources Things that people use and that are found in the natural world, such as wood and water
Nazi The regime in power in Germany from 1933 to 1945, of which Adolf Hitler was leader
Nobel Prize An award presented each year for great contributions to an area of work, such as science or music
pharmaceutical Relating to the making of medicine
PhD The highest degree in any discipline at a university or college
physics A science that deals with matter and energy and their effects on each other
protest To speak out against something
radar A system using radio waves to find objects
radio astronomy The science dealing with radio waves from space
religious ceremonies Special events related to religion, such as festivals or funerals

reproduce To create offspring
resign To give up a job or position
segregation Keeping a group of people separate from another group, based on racial or religious ideas
seismic waves Vibrations, or shock waves, that travel through Earth and across its surface
societies Groups of people who meet to discuss an interest, such as art or music
solar bursts Bursts of energy from the surface of the Sun
species A group of animals or plants that are closely related
survey Measuring and recording details of an area of land
traditional A way of thinking or behaving that is expected or has been done for a long time
tuition Money that is paid to a school in order to study there
vaccine A substance that protects against a particular disease
veterinary Related to the medical treatment of animals
vibrations Quick movements from side-to-side
virus A small particle that causes disease and can spread from one organism to another

LEARNING MORE

Read more about women who broke the rules and changed the world!

Books

Dickmann, Nancy. *Women Scientists in Astronomy and Space*. (Superwomen in STEM). Gareth Stevens Publishing, 2018.

Ignotofsky, Rachel. *Women in Science: 50 Fearless Pioneers Who Changed the World*. Ten Speed Press, 2016.

Maggs, Sam. *Wonder Women: 25 Innovators, Inventors, and Trailblazers Who Changed History*. Quirk Books, 2016.

Websites

Find out more about amazing American women scientists at:
www.acs.org/content/acs/en/education/whatischemistry/women-scientists.html

Read about Hypatia, Rosalind Franklin, and other great female scientists at:
www.britannica.com/explore/100women/profiles

Read about women at NASA today:
www.nasa.gov/stem/womenstem/women-in-science.html

INDEX

ABOUT THE AUTHOR

Cynthia O'Brien has written about many things, from geography and science to animals and history. Working on this book was especially great because she could find out and write stories about amazing women in history. All of these women were determined to be scientists who made a difference—and they did.